THE PASSION OF THE NIGGA

BY: J. WEISSMAN

BOOK I: A PRELUDE TO SUFFERING

Laurel Ave. Publishing™
11600 E. Washington St
P.O. Box 29083
Indianapolis, IN 46229

DEDICATION

Dedicated to Chocolate Chinese & The Brothers: "I've always loved you. I will always love you. Please don't forget me."

Also dedicated to Bubba, Baby-Baby, and Chai: "You rescued me in those days of dark and distant loneliness - Beautiful life indeed."

Lastly, to Ms. Jane's daughter: "Why in the fuck did you lie about me? Why did you destroy our family?

In the end, I would like to thank you for the pain and the suffering. If it weren't for you, this book would be nothing more than a few lines scribbled on notebook paper tucked away in a closet somewhere.

With gratitude, I wish only misery and suffering upon you as recompense for what you did to me and for what you did to our children.

We were a family God damn it! Imperfect and flawed, we were a mother fucking family!

It is my prayer that those, among whom you have hidden, see you for who the fuck you are and that the God of Israel smites you with a vengeful blow. You hurt me. It will take a long time to heal.

I will always hate you."

THE PASSION OF THE NIGGA

A WORD FROM THE AUTHOR

Like Alice in Wonderland, we have fallen into the dark and boundless expanses of Mystery Babylon. Captives; my forbearers were delivered forth, through the Door of No Return, across the Middle Passage and into the Magic Queendom.

Kings no more, we cried aloud to the God we betrayed, and he heard us not. We searched the depths of our captive minds and found our historical slates wiped clean.

Our eyes have seen the glory of the coming of the Lord. But, in the bloody mire of the greatest nation on earth, the greatest nation first birthed combs the scriptures for their native worth.

THE PASSION OF THE NIGGA

And I? I am a child among men. I am a mere dwarf among the intellectual giants of lore. Armed with only my Passion, I navigate a course towards my inevitable end. These words tell the tale of a part of my journey.

THE PASSION OF THE NIGGA

PROLOGUE

I hate to write. It took nearly two years to complete this book. "The Passion of the Nigga" is more than a book of poetry or a book of sayings. It is a container of my fears and hatred. Indeed, it is a fragmented segment of my life. Please don't let the title offend you. "The Passion of the Nigga" is synonymous with the Passion of the Christ. No. I am not calling myself Jesus Christ.

Like Jesus Christ, I was crucified for the sins of others. I was mocked and ridiculed for being in the world but not of the world. I was betrayed by someone close to me. I was delivered over to the Romans and tried according to their laws. My own family watched as I cried out, 'Eli, Eli, lama sabachthani?"

The heavens and all its hosts were silent; gazing down upon me in a kind of stoic apathy that hastened my descent into madness. There I was, in the Delta, bemoaning the day I was born; begging death to swallow me up.

It was there, when I had cried all the tears I could cry, that a voice from within said, "Nigga! Write this shit down. It's the only way you're going to get passed it. It's the only way they will remember what happened."

Let us begin.

PASSION OF THE NIGGA DEFINED

"The Passion of the Nigga" simply means the suffering, agony, and martyrdom of a prophet and teacher. It is, however, much more than this. It is the suffering of a son, the agony of a father and the martyrdom of a husband. "The Passion of the Nigga" is the suffering of a child of God.

THE PASSION OF THE NIGGA

THE PASSION OF THE NIGGA

THE "I" VERSE

I am no one. Set adrift in the endless stream of time, I am a mortal man. I've seen with eyes and have seen nothing. I've seen the sun rise and set miles above the earth yet remained in spiritual darkness.

I've seen foolish men exalted to kingship and visionless men lead many astray. I've seen wise men assaulted by the wicked and wicked men praised as Saviors. I've walked with Giants and crawled with Ants. I've soared with Eagles and eaten dust with Worms.

I've been loved and hated in the same day by the same woman. I've been misquoted, misunderstood and misjudged. Possessing

answers, none have asked. Having questions, none have answered.

In finding myself, I found the God of Israel. In finding the God of Israel, I found the truth. In finding the truth, I found freedom. In finding freedom, I became a stranger to family and family to strangers.

I am, and will always be, imperfect. I have and will always strive for perfection. Eastward, I turn towards Zion and cry aloud to the Great King.

"Dear God, can you hear me?"

THE PASSION OF THE NIGGA

THIS WAS THE END

I was betrayed by the woman I loved and sentenced to imprisonment somewhere in the depths of Mississippi. Crying out to whoever could hear, I was hemorrhaging life as my hair grayed, and hope began to fade.

There I was, standing face-to-face with the end as the taste of metal and gun oil teased my sensitive palate. Her embrace, ever tightening, Death wept as she reluctantly prepared to escort me beyond the veil. In a sudden rush, my index finger contracted, the trigger flew back, and the hammer fell.

"CLICK"

I opened my eyes to find Death huddled in a distant corner, her mascara-lined face buried deep in her hands. It was then that I

knew that she wasn't ready for me and I wasn't ready for her. I pulled the barrel out of my mouth, unloaded the clip and retrieved the dud.

This was the end.

EMO

Like a pendulum set to the beat of "self," she paces her marathon through life. Finding blame in all but herself, she is sublime to all who feed her desire to be more than that skinny dark-skinned girl from La Vergne.

An emotional burden, she defecates on the sanctuary floors of the men who've cared enough to look beyond the monster she portrays. Some days, she's quite bearable while, on others, she's too much to handle.

Followed closely by rumors of promiscuity and scandal, she wipes her muddy feet on the doormats of those who show her compassion. Predators of her innocence, she portrays imperfect men and their well-intended deeds.

THE PASSION OF THE NIGGA

Tears in her eyes, she can't remember when a sunny day in June was just a sunny day in June. She doesn't understand that an act of random kindness was just an act of random kindness. Emotionally destroyed, her life is a straw basket filled with endless regret, doubt, mistrust, and rage.

I met her one Shabbat on Bell Factory road.

THE PASSION OF THE NIGGA

SNAPDRAGON

I've looked into the eyes of an ebony dream and realized a nightmare. As naked as Morpheus was when he stood before the masses, hours before the machines came down crashing. "Zion Hear Me!"

I have a story to tell. I have a future to live and a past to decipher. Researching my historical DNA, I place the pieces of my life together like those puzzles me and Momma used to do back in the 70's. Steadily, I navigate the murky expanses of my destiny as lost as Cristóbal Colón was when he arrived in America.

Her name is Cindi.

Equal parts Passion and Poison, her femininity clings to the Chosen; delivering

them into a state of psychopathic ecstasy that maddens the mind enough to drive one to the brink of insanity.

Deep inside, she's still that little Black girl from Tennessee. Those with eyes to see, watch cautiously as her make-believe persona collapses in on her inner child.

Her tears vanishing into the dark expanses of an emotional abyss, she hides her face to deal with the pain of reality and turns within to embrace the constructs of fantasy. A Snapdragon, brilliant and alluring, she is planted somewhere distant and deep in creation.

She was there when Beatrice wasn't.

THE PASSION OF THE NIGGA

HUMAN BEING: PART I

I'm a Human being; an early morning surprise to my Momma and Daddy, I was prophesized to come forth, from the womb, in the month of the Bull. Growing happier day by day, they watched as Momma's tummy got full and swooned at the idea of two becoming three. Good thoughts they fathomed of me as they surrendered themselves for the sake of their growing family. I was born on a Saturday.

I was their living mausoleum. Into me, they poured all their hopes for tomorrow minus the dreadful things upon which they had fallen in the years before. Placing me high on their shoulders, they showed me more than what they had seen in their lives. It was Life; new and full of promise. Behold! Humanity begins at home.

THE PASSION OF THE NIGGA

THE PASSION OF THE NIGGA

SEE NO EVIL

I've experienced "Evil" and suffered the slanderous kiss of the "Devil." Wrapped in mahogany hue, she calls upon the God of Israel with one tongue while spitting curses with the other.

She is the reflection of her mother; bending truth to manifest the reality of those for whom she preys. On those for whom she prays, calamity metastasizes. Her reprobate radiance rises in the East and sets in the West. Only those who see her masks know her best.

The truth she despises as she chooses the best rocks to throw into the windows of your inner peace. Her salaciousness fails to cease as she greets the world in a Hebraic

tongue while tearing down her home with her own hands. Sunday's daughter is a Daughter of Zion as it suits her.

THE PASSION OF THE NIGGA

INIQUITIES

I awoke drowning in a pool of iniquities. Facing the heavens, I watched as the promise of a brighter tomorrow darkened. Clouds began to block the sun and the shit precipitated as "I told you so" sprang forth from the sun-scorched soil of my devastated landscape.

As I watched in disbelief, my "Biologicals" turned away from the awful sight of my miscarriage from Lady Liberty and prayed to the god of their oppressors as I was passed along by those who would devour my soul. With Biblical hatred, I watched as Mystery Babylon fed the Children of Israel with the blackened blood of dead things; her

colostrum fattening the souls of those latched to her musty teats.

Set adrift in this "American Dream," I am invincible before the "menace of the years." In the presence of the God of Israel, I was his only begotten son in whom no one believed. I was an astonishment to the nations as my temple was destroyed for the sake of feelings, want of financial gain and for the recompense of misplaced trust.

CINDI

Cindi ain't no Gold Digger. I say this because Gold appreciates slower than Silver. So, her eyes are fixed on what appreciates the quickest. Those thighs the thickest, she rocks her hips hypnotically singing weaker Niggas into a state of slumber from which they awaken completely used. But, she ain't no Gold Digger.

Seeking securities like stock brokers, she builds with Cats that deal in Futures then compounds her interests. Changing clothes quicker than politicians change opinions, she can be your best friend. But, when those Futures weigh less than the collateral damage of her past, she's your enemy then.

Fashioning herself as both the Pimp and Pimped, she cries out, "Bitch, where my money?" and "You're just like the rest of those Niggas who abused me." She's got an Uzi for a mouth and a temper that levels whole relationships right before her eyes. Her lies are the kind she tells herself.

Unable to see the monster within, she blames everyone else for the shit she done stepped in. But, there's a sucker born every minute and a well-intentioned Captain Save-A-Ho born every day. And those Futures continue to appreciate.

THE PASSION OF THE NIGGA

BEGINNINGS AND ENDS

Everything that has a beginning has an end. I remember the beginning, the lucid radiance of the days of old. I remember that steaming hot bowl of oatmeal and Momma when she used to spread it around until it was cool enough to eat.

I remember my Daddy's face, when he would walk through the front door after work and how his face instantly illuminated at the site of his wife and children. We were his heaven, and he was our king.

I remember Cassandra and how she used to smile at me. She had the whitest teeth. I knew, when I met her, that she was my soulmate. She was my comfort as I transitioned into adulthood and I was her hope. I remember

how her smile darkened after the abortion and how the love I once felt for her muddied into an ever-increasing hatred. She gave me a son. Then, she took him away from me.

I remember Ms. Jane's daughter sitting with Pam. O, how I beheld her beauty as ignorantly as a moth drawn to the warm glow of a distant flame. She was a bridge over the hurt and certain involuntary events that ushered me from the shame of Casandra into a life that mimicked that of Daddy's.

Ms. Jane's daughter loved me less than she loved the exposure I gave her. When my well dried up, she fashioned her mystery of iniquity and enveloped everyone and everything I loved within it. Then, she became Casandra and took my children from me.

This prose is a testament to life. It is evidence in the absence of compassion. In the wake of Ms. Jane's daughter, there was no more natural love or romanticism with which I

could spend on the Daughters of Zion. Dutiful and loveless, I endlessly toiled in the cotton fields of my spiritual Mississippi in search of enough compassion to save myself from taking my own life.

THE PASSION OF THE NIGGA

TWO FACES

Sunday's daughter has two faces. One face faces the world. The other face faces me. One face is radiant and kind. The other face is evil as shit.

There is the face that complains about inviting one of her oldest sisters into her home, a face that dares to cry foul because her sister knows how to be a good wife and does it. There is a face that would rather throw shade on her sister for being a good wife than try to be a good wife herself.

To spare the world from the drama, she unleashes havoc on those closest to her adorned with masks made of the kind of lies she tells herself.

Real eyes realize real lies.
Real eyes see Sunday's daughter.

THE PASSION OF THE NIGGA

THE PASSION OF THE NIGGA

UNDER THE SUN: PART I

I've seen these things under the sun; deception dipped in familiarity; a pleasant appearance, though nebulous by nature. I've seen brothers hate their sisters and sisters hate their brothers. Between them are fake smiles and semi-cordial pats on the back. No one sees where the other is coming from because there are so many hidden agendas.

Treading Ockham's Razor, they march victoriously towards their inevitable demise. Yielding to the ebb and flow of causality, their razor tore feet trample the alabaster floors of the holiest of holy tempting Beasts who savor the scent of their decaying souls. But, alas, the God of Israel witnesses all things under the sun.

THE PASSION OF THE NIGGA

THE PASSION OF THE NIGGA

WHO AM I?

I am considered a peculiar individual. Having sampled many of the world's cultures, my personality was influenced by more than what was relevant to those who may consider themselves like me in some way or another.

I've seen places likes Scottsboro, Alabama before Black folks were allowed to cross the railroad tracks. I've witnessed the sunrise over the Smoky Mountains and held on to its last rays, at Newport Beach, as it dipped below the horizon.

I've lived in places like Greenwood, Madison, Sewell, Washington DC, Ballwin, and Stone Mountain. I've loved once but was too young to negotiate it. I've lived twice but was too old to appreciate it.

I've seen the passing of a close friend and witnessed the birth of a new life. I've shoveled shit at Negril, for a semester, to afford a trip to Hampton only to arrive at Hampton and realize that true love is letting go and moving on.

I've sported everything from an Afro to a Bald-Head but didn't find myself until I was Dreaded. And, when I was Dreaded, my appearance was indeed dreaded; frowned upon by strangers I never knew and reviled by "familiar" strangers I thought I knew.

I am the product of the struggle; brought forth just a few years after Blacks and Whites were allowed to piss in the same restrooms and learn in the same classrooms.

I am the result of the Hustle, brought up when Pork-n-Beans and Hot Dogs were an everyday meal and picking Dandelions for Momma, outside Winchester Apartments,

before that ride to head-start, was love that was real.

I am a Nigga because I am one of a few to whom this word's truest meaning applies. I am the product of all the above, a witness of the events that are presently unfolding and the catalyst for that which is to come.

I am me.

THE PASSION OF THE NIGGA

THE PASSION OF THE NIGGA

TEMPLE OF ISHTAR

I sent a message to Ms. Jane's daughter apologizing for being such a poor ass husband, but I got no response in return. Typically, her silence wouldn't mean much but, in this instance, it did because I was trying to bridge a gap that I didn't create on my own.

Bitch sat there, in the temple of Ishtar, declaring herself the conqueror of the mountains of Zion while prostrating herself to the Queen of Heaven.

Again, and again, I retransmitted radio signals into interstellar space apologizing for shit I didn't do. My response was silence. There, I stood trying to be that God-fearing, righteous, "Black" man when that Bitch

deserved an ass beating. But, that's not the type of Nigga Daddy raised.

QUIN
(A TRUE DELTA TALE)

I saw this sister at the corner store one evening on my way back home from Gorillaville. Standing just in front of me, I complimented her. "Excuse me," I said. "Please don't take offense, but you are one of the most beautiful women I've seen since I've been in the Delta. What's your name?"

She scoffed at me, pulling her snotty-nosed daughter close. "Who the fuck do you think you are?" she asked. "I don't need no God damn Nigga for nothing. I don't need your compliments either. Bitch ass mother fuckers are always trying to fuck. Bitch ass Niggas are always trying to get what I got.

Mother fucker, I don't need shit from you but for you to leave me alone."

That was some scary shit if I must say so myself. I remember when a compliment was just a compliment and nothing else. Right about then, the debit card reader beeped. The Arab behind the glass told her, "Ma'am, it says your EBT card was declined."

"What?" she screamed. "How the fuck am I supposed to eat?" "Am I supposed to put all this shit back because your debit card reader is broke?"

The Arab repeated himself. "Ma'am, it says your EBT card was declined."

She looked back at me and said, "Pay for this shit, and you can get it." I was intrigued. A minute ago, I was a 'God damn Nigga' and a 'Bitch ass mother fucker.' Now, as if by some miraculous turn of events, I was the brother who would pay for two bags of chips, a pack of Swisher Sweets and a two-liter RC Cola for some ass.

I reached into my pocket, pulled out a ten-dollar bill, and passed it to the Arab. He gave me back change. I gave it to the sister. She asked, "What's this for?"

"Humility," I responded. "I heard you could put that shit on layaway over at Best-Value." I don't think she understood what I meant because she just stood there like a deer in headlights. She stepped aside; I paid for what I was purchasing and exited the store.

"Hey!" she said, following me out into the parking lot. "My name is Quin. My friends call me 'Big Red.'

"Nice to meet you, Quin," I replied.

"Here is my number," she said as she passed me a torn off piece of paper. "I don't like owing people money."

THE PASSION OF THE NIGGA

SOME PEOPLE: PART I

Some people are dreamers. Swimming in an endless stream of fantasy like fish in the deep blue, so occupied with surrealism that they can't even see you.

"Row, Row, Row your boat, gently down the stream.
Merrily, merrily, merrily, merrily.
Life is but a dream."

Some people are asleep; walking and talking as if they're awake. Yeah, I hear the words they use though most are fake. They're usually too busy walking crooked to see straight. Designers of they own fate, they never realize they've been deceived.

Some people lie; whether outright or through the tears they cry. They can't seem to tell the truth. On swords, they lie; dodging

what they said yesterday with more magnificent illusions today.

"Liar, liar pants on fire."

Some people are depressed; too wound up within to redress old wounds. Like show tunes, the story of their life seems to replay in their minds. And, the Depression Deepens.

Some people want to be love but can't love in return. A thing they yearn, they ain't never really had it in their lives. So, how could they recognize it when it arrives?

Searching deep inside, they only succeed in finding well-intended feelings voided. Their emotions toyed with; some people could give a shit.

Some people don't deserve happiness. Too high on the smoke they blow up their ass, they sit in the special needs section of the School of Life thinking their shit is the only shit that smells right. In reality, their life's a

living Hell, and they're in the worst condition of all.

Some people…

THE PASSION OF THE NIGGA

THE PASSION OF THE NIGGA

HUMAN BEING: PART II

Peering through the windows of her soul, I searched for the love I once found in Ms. Jane's daughter.

I'm a Human Being the only one fighting to save a marriage. And, as I searched desperately for some sign of life beyond the windows of her soul, all I could see was the disappointed faces of a girl child and her two brothers; Nothing else.

This is the Passion of the Husband. This is the Passion of the Father. This is the Passion of a Son who needed his Momma but didn't know how to say, "Momma, I need you to help me through this." This is the Passion of the Nigga.

THE PASSION OF THE NIGGA

Have you ever seen such a thing? Have you ever traversed the sands of time carrying entire realities on your shoulders only to find that you were nothing to everyone you loved?

I see the Delta, and she greeted me with open arms; soothing me with her genteel hospitality and promises of second chances, but I secretly hated her for being so damn small minded.

Staring frantically through the windows of Ms. Jane's daughter's soul, I found nothing but Alienation of Affection and a subtle form of contempt that had been there since we both said, "I do" fifteen years before.

Though the Delta loved me unconditionally, I pined obsessively over the thought of being a failure, and I was alone; so alone that the sultry fragrances of Mississippi women began to intoxicate my even-keeled constitution and I was weakening.

The windows of Ms. Jane's daughter's soul were shut. I could see no more light in them. Ominous darkness was before me.

In the end, I couldn't see anything anymore. I couldn't see her. I couldn't see what she was about to do to our family. I couldn't see my children.

THE PASSION OF THE NIGGA

THE PASSION OF THE NIGGA

UNDER THE SUN: PART II

I've seen these things under the sun; men of valor who strive for perfection though they're imperfect. Though these men are neither proud nor boastful, there are those who lay in wait choosing the right moment to assassinate their character.

I've seen good men made filthy by perception; private men make exceptions to privacy and regret it. Giving favor to the faux, the faux take that shit for granted and abuse their kindness. All the while, these men are systematically destroyed by the same spirit moving among the weak. But, the weak are unaware that they're being used as weapons of mass destruction.

THE PASSION OF THE NIGGA

I've seen division called unity and unity become a reason to divide. I've seen men and women, alike, full of pride. Then, they fell.

Under the sun, the God of Israel watches patiently as the time of the Gentiles ends and the era of Jacob begins. But, Jacob slumbers; inebriated from wine drank from the cup of Babylon's iniquities. They are, simultaneously, the source of their curse and the catalyst of their redemption; if only they were to awaken…under the sun.

THE MEEK

Disheveled and worn, he is humbled by the world. Stretching forth his hands, in supplication, the sons of men disregard him as they pass. He beholds giants. But, to them, he is invisible.

There is unhumbled magnificence in their eyes. There is no fault or fear or concern or empathy. They're a widow to none.

Hands stretch forth in supplication; his blackened face paints a tale of hardship beyond the horizon of his memories. A foreigner in a land of Dreams, he is awakened to that which the sons of men cannot see. His hands stretched forth in supplication, he speaks.

"Halleluyah."

THE PASSION OF THE NIGGA

THE PASSION OF THE NIGGA

AN ODE TO BETRAYAL

As best experienced by a person betrayed, emotions stayed and festered until one hates or truly understands. But, that understanding is a subjective or mental edifice that suffers only the subject and does absolutely no justice to the object which was initially misunderstood.

Apropos the inner reflection of one's residual image, also known as self-identity and or pride, subtle nuances during an objective emotional session will more than likely erupt in various strata of doubt, suspicion and overall mistrust. This is also a sign of either intended or enacted upon infidelity.

But, I digress.

THE PASSION OF THE NIGGA

My Momma used to say, "a hit dog will holler first". As allegorical as the statement above seems, it is inherently true.

But, that which may seem inherent is indeed a learned affirmation because it requires a level of self-awareness that causes the reflector to be more than one who experiences a sequence of events, but rather one who witnesses the selfsame event as if he or she were watching a movie playing out before them.

Who among us will attest to the fragile span of sociological DNA that encodes for, among many things, the willingness to love and let love?

We move from betrayal and consider it a side effect of the overall journey; an inevitable "down" which is complementary to the "up" we all love to experience; that thing called bliss.

THE PASSION OF THE NIGGA

And, as bliss spirals into deep affection, many practitioners of this ship of relations fail to recognize the gravity of each moment as the consequences of letting go begins to feel natural. The higher one feels, the further away from self-control and individuality one distances him or herself.

From the recompense of that blissful feeling comes the inevitable descent into familiarity. You know, that feeling one gets when a simple hug, soft caress along the upper arm or a kiss on the cheek no longer ignite the fire within.

We rest upon this most assuredly because most assuredly this is the definitive beginning of the end; when the livid familiarity of that which is taken for granted begins to ferment in the emotional sinuses of one, the other, or both.

Shortly afterward, every small thing becomes a tall thing. Every skinny thing

becomes a fat thing. And, every White lie becomes a, *"Where were you when I called you on your cell phone? You must've been with that Creole Bitch Simone!"* But that, of course, is another story altogether.

THE PASSION OF THE NIGGA

SOME PEOPLE: PART II
(THE RE-VISITATION)

Packed in mega churches, hands clasped beneath Gentile steeples, some people beg White Jesus to save them; never fully grasping the number of LEGIONS he gave them. And, the church says "Amen."

Like the Marine Corp, Babylon seeks a few good men to administer the Blue pill to the masses while singing lullabies in the form of scripture talk in Wednesday night Bible classes.

Some people are conflicted with "Christ Insanity." Clouded by what Preacher preach, their spiritual eyes can't see. Crowded hip-to-hip on hard wooden pews,

they're religiously subdued. "Dear God, can you hear me?" And, the church says "Amen."

Some people want to be free, but they're too afraid to leave the plantation. Some people have been given the world because they sold out. Receipt of sale in hand, they believe they're better men. But, no matter how they look or what they know, some people are still Niggers.

Some people are graduates of Affirmative Action. A school of thought that gave them one thing and took from their children the same. They've made it too far to see it. And, even if one were able to show them, they'd refuse to believe it.

Some people are already dead. They just haven't gone down to the grave. Some people are alive. They simply haven't come up from the grave. Some people are moments away from liberation but fail to see it.

THE PASSION OF THE NIGGA

Making bricks with no hay, $7.92 an hour, every damn day, some people deal with chemicals to make the pain go away. Stocking Pine Sol on Isles 8 between 10:00 pm and 7:00am, they smoke trees to get a glimpse of heaven whenever they can spare a dime. To some people, the misery of life is transformed into a more colorful landscape when they're high.

THE PASSION OF THE NIGGA

LOVE

What is love? What is love that it hurts so much? What is it that causes a father to hate the day he conceived his first child?

What is love that it causes a woman's passion to grow cold towards her husband; seeking natural affection through unnatural means as intimacy with Romance novels eventually substitutes the natural feel of a warm embrace?

What is love that it causes a loveless woman to pray for it, get it, then refuse it because it doesn't fit her preconceived perception of how it looks and feels?

What is love that it causes a woman to choose the security of a 9-to-5 over that of her duties as a wife and mother; opting to spend

time with her man as long as Roger permits it? Slowly transforming passion into compassion, what is love that it causes best friends to become friends at best?

What is love that a father can view his baby sons as a casualty of marital bliss; victims of quiet weapons deployed in silent wars?

"I'm turning out to be just like Momma…having all these kids back to back. I never really wanted to be a mother or a wife."

Then, came strife hidden behind well-placed smiles and even better-placed displays of happiness. And, that was that. The world is a stage, and everyone's an actor. But, God bless the Nigga who has his own.

THE PASSION OF THE NIGGA

THE PRODUCT

I was a product; packaged as consumable goods and marketed to society as a "non-violent African-American conformist." A product of my environment, I am only thirty years from retirement as far as I can see. I am so twisted up in the "American Dream", I ain't gone never wash completely clean.

Hiding beneath the skirt of Lady Liberty, I watched as she raised her right hand towards Jerusalem while holding a book of her deeds in the other. Pushed from a living womb, this Bitch was my second mother.

I gave her my essence when, in return, she gave me Michael Brown, Eric Garner, Tamir Rice, Christian Taylor, Alton B.

THE PASSION OF THE NIGGA

Sterling, Keith Lamont Scott, Freddie Gray, Sandra Bland, and Philando Castile.

When I was blind, she made love to me. When I was poor, she fed me. But, when I woke up, she turned on me. Falling from seven figures to minimum wage in less than two years, I feared she would succeed in killing me.

She was sick. Lady Liberty caught that Nasty Woman's dis-ease from sucking every nation's dick, and she was wasting away.

A widow to none, save the God of Israel, she fucked with every nation causing those who poured into her to reek of her iniquities. In the name of Manifest Destiny, she spread her legs as far North as Ontario and as far South as El Segundo, into Mexico and South America.

News of her sickness caused mass hysteria as her suitors began to pleasure themselves and pull out before it was too late.

These lovers she'd hate as strange bedfellows became threats to her Security.

As times before, she'd tap into her sureties for a few more dollars to deploy a few more naval vessels and employ a few more middle-eastern scholars to convince the masses that she's been violated.

All the while her coffers were being evacuated as her best friends reaped dividends and her children begged for jobs. In her depraved socio-economic mind-state, she'd sob, "Why give them jobs when they can have Section 8?"

Intoxicated from her pride, her haughtiness quaked as she longed for more international intercourse. Too proud to give her captives a divorce, she simply writes and re-writes her story and calls it history.

She used to be my mother.

THE PASSION OF THE NIGGA

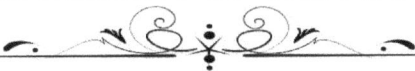

PAULA'S SATAN: PART I.
(AN ACTUAL E-MAIL)

"I'M WRITING YOU BECAUSE WHEN EVIL COMES IN THE MIST, YOU MUST ADDRESS EVIL, I CAN'T BELIEVE THAT YOU WOULD GO AMONG THE HEBREWS AND TELL LIES ON PEOPLE WHO HAVE BEEN NOTHING BUT NICE TO YOU AND YOUR FAMILY.

YOU CAME INTO OUR HOME AND TREATED IT LIKE IT WAS YOURS, AND WE HAD NO PROBLEM WITH THAT. BUT, TO KNOW THAT ALL YOU WAS DOING DURING THAT TIME WAS CAUSING DIVISION AMONG US, HOW DARE YOU GO BACK AND TELL THEM THAT WE KNEW ABOUT THE PREACHA MAN BEING CALLED BY JESUS TO PREACH THE GOD-SPELL BEFORE THE DAY THAT HE MADE THAT ANNOUNCEMENT.

THE PASSION OF THE NIGGA

YOU KNOW, AS WELL AS I DO, THAT HE TOLD US THAT SAME MORNING RIGHT BEFORE HE WENT TO THE TABERNACLE TO STAND BEFORE THE ELDERS. AND THEN FOR YOU TO TELL THEM THAT WE HATED JEWS WAS YET ANOTHER LIE. MANY OF US IN THE SOUTH DON'T UNDERSTAND JUDAISM, BUT WE DON'T HATE JEWISH PEOPLE.

I DON'T UNDERSTAND WHAT YOUR MOTIVES WERE BEHIND WHY YOU DID WHAT YOU DID BUT KNOW THIS; NO ONE NEEDS YOU TO TRANSLATE ANYTHING WE SAY OR DO.

IF YOU HAD A PROBLEM WITH THE PRECHA MAN'S TOPIC THAT HE SPOKE ON OR THE REMARKS THAT I MADE DURNING THAT LESSON THEN BEING A MAN YOU SHOULD HAVE ADDRESSED THAT THEN, NOT RUN AND TELL THE ELDERS WHAT YOU THOUGHT HE SAID.

SO, YOU SEE WHY I MUST CALL YOU ON ALL THE LIES THAT YOU TOLD.

A SPY WAS IN THE CAMP, AND YAH HAS SHOWN YOU FOR WHO YOU ARE, AND I PRAY THAT THE ELDERS WILL FIND OUT ABOUT YOU AS WELL BEFORE YOU CAN START ANY MORE PROBLEMS.

I NEVER TRUSTED YOU FROM THE BEGINNING, AND NOW I KNOW WHY.

THE SPIRIT DOESN'T LIE.

{EVIL} YOUR NAME SHOULD NOT BE JUDAH WEISSMAN BUT" SATAN." YOU ARE ABOUT ONE OF THE MOST EVIL PEOPLE I HAVE EVER COME IN CONTACT WITH.

YOU WERE EVIL IN THE MIST."

THE PASSION OF THE NIGGA

THE PASSION OF THE NIGGA

I AM ETERNITY

I AM Eternity. Hued from the blackest expanses of the Universe, I've formed galaxies, created nations, and destroyed worlds. Civilizations have worshiped me.

I was Zeus to the Greeks, Buddha to the Chinese, Krishna to the Indians, Apollo to the Romans, Thor to the Celts, Waken Tatanka to the Native Americans, Quetzalcoatl to the Mayan and Jesus to the Jesus People. I am Adam, Enoch, Noah, Abraham, Isaac, Jacob, King David, Shlomo and the prophet Elijah.

My ebony skin tells tales of the suns I've kissed. In my mind's eye are the blueprints of creation. I bring order to chaos. I am the finder of the lost souls and father of the Earth.

THE PASSION OF THE NIGGA

I am the teacher of men. Greatness remains in my wake. My praises are engraved in cuneiform. My achievements are painted on the walls of the Pyramids at Teotihuacán. I am the beginning and the end of all things.

Taken beyond the Door of No Return, I've been robbed of my heritage and scattered to the far reaches of the Earth. I was in Alabama eons before the advent of the European.

I was mis-educated, deceived and made proud to be an African-American. I was a fatherless child, a Bastard among nations and a lost generation.

I am the Future, the Past and the "Now." Born of the union of mother earth and cosmic seed, I am "los gentes en dios". I am the God of Israel's firstborn son. I am an immortal man dwelling in a perishable form.

I AM THAT I AM.

THE PASSION OF THE NIGGA

AGAIN

A Brown child is born and is loved by his Momma. He is in the image and likeness of his father, but he struggles to be a better man. He lives for himself but is hated by the world.

He is intellectually pervasive, but his Blackness paints him ignorantly against the tapestry of Western European ideology.

Having been shown no way but poverty and prison, he is lost. Being lost, he finds himself. Though he finds himself, he is alone. The loneliness begets anger. Anger begets rage. Now, in the darkness of the Delta, he is forced to begin;

Again.

THE PASSION OF THE NIGGA

THE PASSION OF THE NIGGA

UNDER THE SUN: PART III

I've seen myself under the sun, full of pride at one moment and laid low the next. I have traveled the world in pursuit of knowledge, but in the end, I was even more vexed. I've seen love translated as dislike because its personage was not ideal, and hatred translated into love because it looked and felt good.

Under the sun, I've known both good and evil. I've walked with wise men and stumbled with fools. I've done righteously to the unrighteous and unrighteously to the righteous.

In all things, I've learned that nothing is new to the God of Israel. At my end, my

flesh shall return to the earth, but my deeds shall echo in eternity, under the sun.

THE PASSION OF THE NIGGA

THE DESERT OF THE "REAL"

I have a Passion. My soul cries out to the God of Israel as I walk through the Valley of the Shadow of Death. The silence aggravates the sanctity of my mental fortitude. Spiritually subdued, I find no purpose in my cosmic journey towards the light. Looking East, I stretch forth and take hold of Israel's hand, then rest upon her bosom.

For, I am bound by this Babylonian yoke and its heavy. Depressing me further down, I tread knee deep in the bloody excrement that spews forth from Lady Liberty's underdressing.

Standing shoulder to shoulder with sleeping giants, I cry out to the furthest reaches of creation. "Dear God, can you hear me?"

THE PASSION OF THE NIGGA

As I march victoriously to my inevitable demise, my head remains lifted. Neither a Saint nor a Sinner, what has been fashioned for me is a path of sharpened razors whereupon either side is an extreme and I am the Mean.

I am your average everyday Nigga traversing a jagged path of sharpened stones and ancient bones, and my feet are bleeding. Too confident in the God of Israel to accept failure, my way is the way of most.

Seeking life in a state of suspended expectation, darkened days greet me. Deep down inside, I know Lady Liberty will eventually defeat me. Still, my head is bloody but unbowed. I am still the master of my fate. I am still the captain of my soul as I traverse this desert wilderness on my own.

THE PASSION OF THE NIGGA

CASSANDRA

I met her at a bus stop in Northwest D.C. on September 22nd back in 1993 - Cassandra. In shades of ebony eroticism, her presence painted a poetic masterpiece of she and me together upon the blank canvass of my mind.

Back then, it seemed like universal forces brought us together, colliding into one another at the speed of life and reckless consequence. I had never seen a woman so beautiful as she sashayed into my life like a dream; a beautiful dream of what might be between us like walks in the park and holding hands on cold December evenings as we stepped lively to catch the Green line to Metro Central.

THE PASSION OF THE NIGGA

A smile that rivaled the rising sun and the eyes of a goddess, I hoped she would give me her telephone number when asked. When asked, she looked up at me and smiled.

I found my soulmate.

ODE TO DEATH

Staring into the eyes of my sons, I've come to grips with my mortality. Searching endlessly for life in the arms of Ms. Jane's daughter, I watched as life sprang forth from her loins and took air.

Years later, she took these lives from me because her mother fucking feelings got hurt. Years later, she took these lives from me because she thought I was financially spent. Years later, she took these lives from me because a devil sold her on the notion of "financially castrating him for hurting you." Years later, she took these lives from me because she had no life on her own without them. Years later, she took these lives from me to destroy their lives and mine.

THE PASSION OF THE NIGGA

O, Death, behold the menace of the years!
O, Death, behold the Destroyer!

Look maliciously upon this blackened soul who deserves a painful death more than she deserved a good life. Look maliciously upon this blackened soul who deserves a painful death more than she deserved leaving Ward. Look maliciously upon this blackened soul who deserves a painful death more than the social benefits for which she traded our family. Spare her not every moment of pain and suffering heaped upon my imperfect soul.

.

COLORFUL FLOWERS

I am overwhelmed by a burden. Like atmospheres of unconscionable air, I bring only anguish. Somewhere deep beyond the superficiality of my misgivings, I resurrect a monument to my Passion. For, we all must carry our cross. I carried her's too.

My blood bleeds red then turns blue as it meanders into the chemistry between that evil Bitch and me. There was a day when I would have given my life in exchange for her's. Now, the only thing of mine I give freely is all my pain and suffering.

I hate her.

Like wildflowers set abloom, blending · into hues of indiscernible catastrophe, she

became nothing more than a moment that had delusions of eternity.

PAULA'S SATAN: PART II
(IN ANGER, HE SPOKE)

Called I was Satan after He Said She Said I was lying looking around wide-eyed spying just because some things were said I said nothing but was given words from remote locations. From diverse persuasions, Satan was the name Paula gave me.

Character assassination brought forth by loose words read across computer screens; turn. For everything, there is a season.

Turn.

Turn to what you assumed and who you presumed because no one never asked the accused have no reprisal just cloaked rivals like Klingon Birds of Prey, they lay and wait in outer space darkness clings to them like

THE PASSION OF THE NIGGA

Melanin on Negroes often never fully appreciate the gift of oneness until its gone 'til tomorrow full of sorrow was the son of Peewee.

When the spirit changed, the game re-arranged leaving this Nigga apart from what he knew he was an experiment of the highest degree who failed to see the fullness of "evility."

Then, so-called Bible Masters baited, burning bridges to places they've never been deceived but, someone in their midst told the tale of two cities. I shed those good tears as wolves in Culture Clothes and Head Wraps burn them with words that are too heavy for them to understand nothing but what they're told.

Paula Behold!

I wondered why it hurts so much to be hated. Gossip and Rumors reiterated. What was destined to be will be in time these rhymes will live long after my last breath leaves me.

Guilty of nothing more than sitting amongst thieves and temple whores often play virtuous behind closed doors after that "nut" got swallowed. Paula's poison is killing them softly tread through the minefield of the psychological tempest of her past, present and future never looked so gloomy. So, they doomed me.

And, the son of Peewee was labeled Satan.

THE PASSION OF THE NIGGA

THE PASSION OF THE NIGGA

MINE EYES BEHELD A DREAM

Mine eyes beheld a dream. Like Martin Luther King, I gazed down from the mountaintop and witnessed delusion spinning forth in all manners of confusion. Too much to take in, its deceptive properties caused even the Elect to behold their own words because their own words were easier to obey.

Failing to take heed to what the "Higher Man" says, they tip-toe along the outskirts of righteousness like illegal aliens along the Rio Grande. They're too self-righteous to accept that they're simply men who're capable of greater wickedness than good.

Mine eyes beheld a dream wrapped in insatiability and discombobulated into a form

that has interwoven itself into the fabric of our lives. Alien and distant, it is second nature to those who cannot see it because they're asleep.

It is the "American Dream."

THE PASSION OF THE NIGGA

PRELUDE TO SUFFERING

The pain is internalized. I look upon myself, from within, in an attempt to bring order to the chaos. Sorting through the wreckage of what has become my life, I seek normality despite the bludgeoning of chance.

Look upon me as the turbulence within twists in an emotional tempest spiraling downwards, downwards, downwards into a deepening state of depression. Becoming the object of my hatred, I succumbed to an ill-fated end before life ever began.

I died in 1994 when my first son was aborted after his mother learned that I would be leaving Washington D.C. She couldn't bear the thought of being a single mother. I was told about my son after the fact because she was

pro-choice. She chose for my child and me because it was her body.

I deferred the shame of telling Daddy that I got a Louisiana girl pregnant and the child was aborted because she refused to be another statistic. Oh, the disbelief and condemnation I circumvented only to fall into a greater Hell one Thursday at Mary B's restaurant.

I was born again in 1998 the moment I beheld Chocolate Chinese being drawn from the living waters and was given strength through the births of my sons a few years later. But, Ms. Jane's daughter was a shrew with little home training beyond a certain degree of cordial civility. Beautiful and petite, soulless and empty, she was a paper wife of tepid moral conviction; impotent and conniving.

It was happiness at a cost. How much are the lives of three children and a 15-year marriage when the love is lost?

THE PASSION OF THE NIGGA

Dreadfulness filled me as true happiness became a commodity I was too broke to afford. The wife I saved was giving no loans to cover our marital shortcomings.

Given homes, cars and exposure to life beyond the confines of a simple place somewhere deep in Alabama, she had nothing to give in exchange but disdain and a dull rusty knife with which she intended to end me at a time of her choosing; "a fox who outfoxed the fox," Momma said.

From Mississippi, I watched as that Bitch changed me from an every day Dad to this Nigga who only spoke to his children four times a year. The God of Israel bore witness to the weeping I wept and the tears I shed down there by the Levees.

In those days, misery became as common as breathing. So, I inhaled the Delta deeply. "The Passion of the Nigga" became my therapy as the words that spewed forth for my

wounded soul metastasized in the form of ink on paper, cushioning my descent into Hell.

GONE FROM ME

I have dreamed a dream, and now, that dream has gone from me. A facade constructed to mask the pain drapes upon me like curtains concealing the stage play of my life.

Wiping the tears away, my darkened face moves between the God of Israel and the world that beckons. A conundrum becomes of me as my rebirth hides me from the sight of those who watched from afar.

The dream I dreamt has become a prison for my soul. Giving to those who are unwilling to give, I am being crucified by those who're living to kill me.

I am a constant gardener.

THE PASSION OF THE NIGGA

I've embarked upon a journey immersed in futility. I am a Prophet treading waist deep in a never-ending cistern of lunacy. My only refuge is the hope that I will be remembered by my children when I am gone from here.

Toiling upon the footstool of the God of Israel, my tears are no more. I've become part of the misery that besets me. I've become the mechanism of my demise. Aloof, in a dream-like reality, I wander, in the wilderness, under the sun.

MS. JANE'S DAUGHTER

From the Delta, I watched as Ms. Jane's daughter spewed forth unintelligible interpretations of a manipulated reality weaved carefully into webs of deceit to entreat the weaker sensibilities of her European Massa.

From high atop the Bench, they leveraged their Roman Curia against her Aborigine ex-husband and swooned at the sight of her fearful tears though they were as fake as poorly written fiction.

Hands drenched with the blood of a child of God, Ms. Jane's daughter was an invitation to those BARflies; enticing them to bruise the right heel of Hoover's grandchild. Humble and contrite, she ascribes ill-fated

verbiage in the direction of a Nigga who loved her to his demise.

There, before the Crown, she stood there lying; supplying codes and coordinates to the Rebel Alliance. In defiance of all that's holy, she turned legal tricks sick enough to make Satan sick while her "word twister" stood idly beside her.

They never tried her to determine whether her truths were the God of Israel's truths or at least truths that were true enough to render righteous adjudication.

She made a folly of the Alabama court system, lessening the value of that Judge's academic career. But, alas, that's what she does; lessen things. "O, what a tangled web we weave when first we practice to deceive!"

Ms. Jane's daughter coveted a way of lies more than the way of wives. Let it be told that her reasons for doing what she did were because she was afraid of polygyny. Let it be

told that a Nigga told her that he was going to fuck Cindi whether she agreed with it or not. Let it be told that a Nigga said his very own girl child was only good enough to be another Nigga's second wife. Nothing was furthest from the truth. But, to the ears of some, the furthest from the truth is the nearest to believable.

Ms. Jane's daughter bottled that shit up, labeled it as lemonade, and sold it to her folks and mine. They drank that shit without question. The Law of Moses was done away with because Paul wrote a few letters about it "...in the name of Jesus; Amen".

The first son was forsaken because he was the first to awaken. His reward was Court ordered Child Support and two tablespoons of beans and rice chased by an eight-ounce glass of water every day until further notice.

Ms. Jane's daughter was a product of selective breeding. She was the product of

Matriarchal sieving; separating the proverbial wheat from the chaff until the perfect Nigga killer was created.

But, all a Nigga could see, back in '95, was a fine ass sister that appeared to be wife material when she was a Temple whore painted to appear righteous. In retrospect, she was only a beautiful nightmare that manifested in a Nigga's real-life scenario.

In the end, I can't blame her for the Hell she brought into my life. Before this all went down, Daddy warned me not to marry her. In as much as this is true, my blood is on my own hands. So, in a Biblical sense, I consider this segment of my life a "Jonah" experience.

DEPRESSION DEEPENS

In the silence, a symphony of tears rains down upon these pages as my inner being poured forth from the gaping wound of my afflicted soul. Gazing through the looking-glass of my life, clarity paints a perfect picture in the aftermath of these calamities.

Clairvoyance in things long past, I grope about in present darkness. Listening for the voice of the God of Israel amidst the confusion and contempt hurled by those who hate me, I gaze through the gates of the Kingdom at those who mock the Almighty in the pride of their iniquities.

I am a child of Zion.

Sorrow is an anchor holding my vessel steady upon the high seas of disillusionment.

THE PASSION OF THE NIGGA

The winds that carry my sails are filled with the ghostly howls of self-abusiveness.

In the end, what will be will be. I will transcend the fates that beset me as I negotiate a meandering path towards an uncertain end.

HATE
PART I

Hate pushes and pulls. It tears apart and builds false alliances. It is painful and maddening. Hate decides fates and redirects destinies. It can exist among relatives and grow like wild grass between Husband and Wife.

Hate can cause a man to consider taking his own life. It is a life stealer, seeking misery and discontent. Hate will never be happy. It can cause disillusionment and warp perceptions.

Hate withholds the truth from children and encourages them to resent their father. It is ignorant, unhearing and does not compromise.

She hated me for fifteen years.

THE PASSION OF THE NIGGA

THE PASSION OF THE NIGGA

A PRAYER FOR THE FUTURE

Father, I tried to kill myself. I placed a bullet in the chamber of a newly bought pistol, put the barrel in my mouth and pulled the trigger. I awoke to find myself still among the dead. I have been a failure in life.

Look upon me Father; a failed son. Look upon me Father; a failed brother. Look upon me Father; a failed Father. Look upon me Father; a failed Husband. And, today Father, I failed to take my own life.

For there is no honor for me. Even Death rejects me. Indeed, they shall find me once more, on the stoop, out back, engulfed in a cloud of purple haze. They shall speak as they pass by and I shall speak back. Look upon me; Father. For, there is no one else left who

can save me. Jesus is dead, Allah is too. All I have is you. Grant me a tomorrow that was better than today.

THE PASSION OF THE NIGGA

ODE TO POLYGYNY

Pieces of women I see as I tread through the ever-welling tide of Black femininity. Made in America, the Daughters of Zion are only Daughters of Zion seen through illuminated screens of social media. Posting and tweeting their virtuous Dogma, they portray a form of piety that's as deadly as hidden assassins.

Mixing and matching certain characteristics and physical appeal, the amalgamation of such is surreal as the perfect woman manifests herself in the form of two or more individual women.

With the multiplicity of individuals, the multiplicity of really ugly traits multiplies

steadily as one attempts to assemble the perfect woman out of scores of broken ones.

Here, is where I find myself.

THE PASSION OF THE NIGGA

WHEN I'M GONE FROM HERE

When I am gone from here, who will bear me upon their shoulders? Who will come forth to sit and weep as I lay before my loved ones? How many tears will be shed by those who truly loved me?

Will my children be there? Will Chocolate Chinese remember me in the days of my youth when she and I used to play "hide and go seek"? Or, will she remember the day her Momma told her I left them for another woman and I was never coming back? Are those mascara drenched tears for me or the satisfaction that she'd waited for this day to come?

Seated beside her mother, will she ever know how I pleaded for them? Will she ever

know how I begged her mother to be the wife Ms. Jane raised her to be? I needed her in the days of my youth, and she turned away.

Will my sons speak kind words of me? Will they remember when we used to play and run and jump and wrestle until we were too tired to move? Will they remember that I am their father?

While sitting there beside their mother, will they remember the man who loved them more than he loved himself? Or, will they remember the man their mother said left them for another woman? I loved her, but she loved the bright colored façade of Babylon.

At night, I used to cry down here in the Delta, curled up and alone, rocking myself to sleep on an inflatable mattress with a hole in it. Sinking, sinking, sinking slowly was I. Thus, was the irony of my life.

To Ms. Jane's daughter; Damn you! Damn you for forsaking the God of Israel.

Damn you for lying to our children. Damn you for not being there when I needed you the most.

When I am gone from here, who will touch my face and kiss my lips one last time? Will those who knew me in life gather to witness my demise with choreographed tears and rehearsed displays of sadness. My lifeless form dressed modestly before them, who will even give a fuck about me or the life I lived?

THE PASSION OF THE NIGGA

THE PASSION OF THE NIGGA

DOWN

Green; sick Green, bleeding into the emotional tapestry of a colorful, vivid me. Like storm clouds building in the distant Delta skies, trouble comes my way.

Birthed into a series of unfortunate events, I am the Invictus. My blood, crimson and warm, pour forth from my inner child, spilling all over the nicely manicured façade of Babylon. I am you. "Dear God, can you hear me?"

These tears run dry at the Pearly Gates as I look beyond and see Granddaddy, Uncle William, Uncle George, Uncle Bay-Be and my unborn son stolen from life by his mother. He was my first seed, planted on an Autumn

evening in the District of Columbia. "Dear God, can you hear me?"

I am met with ridicule and disdain as I turn to face the world. Those clouds in the distance are building ever more; giving rise to darkness and all the things that hide in its deep places.

Holding fast to the last rays of sunlight, its amber radiance thinning, I take a final breath and close my eyes; trading the darkness that envelops me with memories of Momma's smile back in the 70's and Daddy when he used to bring Pasquale's Pizza home in that brown paper bag.

But, these memories are as old as the 70's, and they have worn thin on the treadmill of my mind because I've used them so much. As I spiral deeper into the darkness, I entreat the God of Israel and pray that Daddy and Momma will forgive me for disappointing them so much.

THE PASSION OF THE NIGGA

Downwards, I fall into the abyss. There is no end to the darkness. O, what a colorful life I've lived; pristine and radiant; vibrant and rich. Yet, the darkness has consumed what is left of me.

THE PASSION OF THE NIGGA

THE PASSION OF THE NIGGA

AT THE LEVEE

In the face of all that besets me, I hold back more tears than the levees hold water down here in the Delta. Reaching the point where I can hold no more, I cringe as the sky darkens and misery begins to build.

Looking within, the pantries of my soul contain no more sunshine. Ms. Jane's daughter promised we would always be together when the days were new, and the skies were blue.

I took her as my wife without ever really knowing her. Momma and Daddy were as shocked as me. It was a life spinning uncontrollably.

Momma used to always say if I was half the man my Daddy was, I'd be alright. So, I shot for mediocrity with hopes of sharing similarities with him. I stayed awake at night searching for the man inside but, he was as distant from me as the God of Israel was all those Sundays at Cornerstone Baptist Church.

In time, Ms. Jane's daughter and I became five. Happiness, though never truly realized, transformed into a dreaded sense of assumed obligation. Down here, in the Delta, I climbed atop this levee to clear my mind as memories of Chocolate Chinese and my sons eventually faded away.

I can remember letters from her during the holidays; how the anticipation would swell to an almost unbearable crescendo as I reached inside the mailbox to find her name to me. These letters spoiled into letters from her Attorney accusing me of abandoning her and my little ones; pushing me ever closer to an eventual mental breakdown.

I needed her, but she was too full of the shit shoveled into the emptiness of her

decaying spirituality to see me. Here I am, pining obsessively over the years wasted and the lies told as the skies darken, and the rain begins to fall.

THE PASSION OF THE NIGGA

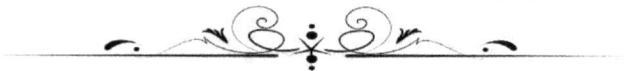

IT WAS ME

Bitter endings are endings all the same. In retrospect, all the crying and emotional fanfare were done for the children; a sequence of planned and choreographed displays portrayed to make my seed grow to hate me.

The memorial of Ms. Jane's daughter's disdain came in the form of a family photo minus me. I looked but failed to see the reason why I was absent. Draped in black, they all gave cheese to the photographer as if trying to show that life goes on after my paternal demise.

I truly hate that little singing girl from Alabama for pretending to be something she had no intention of ever truly being. But, I feel for her because she had something good.

THE PASSION OF THE NIGGA

When she was too fed up with school and dropped out, it was me who dropped in. When she was too nervous to stand before a bunch of mother fucking ten-year-olds, on the first day of school, it was my words that held her above the murky depths of her ruined self-esteem. She learned to draw strength from a Nigga she would one day betray.

THE PASSION OF THE NIGGA

TEMPLE HO

Everybody knew her; a collage of beauty soiled in the proverbial cistern of the minds of brothers she would greet. A few kind words would land her on your lap; passing herself around like Martin Luther King Jr. fans at Sunday Service. Wading deep in the filthy thoughts of Niggas who've known her, she struggled to convince herself that she was more than a fat ass.

Gracing the people of the Tabernacle with superficial elegance, she would take her seat, inconspicuously, and turn to the Book of Niggalations 15: 1-29.

Let us read together.

Beyond the shallow gate of her rehearsed demeanor, she quietly made peace

with the truth of never becoming anyone's wife or any child's mother.

She wept though her eyes could shed no more tears. Broken and scarred beyond beneficial use, she punished herself with self-abuse as she led the righteous passed her gates and into her Temple.

THE PASSION OF THE NIGGA

MR. DEVIL

Mr. Devil, I see you at work and church. I see you at the Market when Sundima and I be shopping and at the schools where little Brown children learn.

I see you looking at me in that Police uniform and from behind White hoods and bed sheets. I see you sprawled out in the sun getting red.

I see you on the television talking about poor ass Niggas like me and rich ass mother fuckers like you. I've even seen you on the moon and at the deepest depths of the deep blue sea.

I see you, Mr. Devil, watching me.

THE PASSION OF THE NIGGA

PATH CROSSERS

We walk the line; facing inner demons at the crossroads somewhere deep in the Delta. I see the God in we as we speak Life into the Death we see.

A prophetic nomad, I AM.
Before Abraham, I AM.
Before the son of Hoover, I AM.
Mississippi Moses, I AM.
Jesus Jr., I AM.

I am just another Nigga to more than a few. Most misconstrue my words because they're unable to unravel the parables I spit. Most can't hear the wisdom that spews forth from my soul as it delivers a testament to the lives I lived.

THE PASSION OF THE NIGGA

To Earth, I give aberrance in the form of word speak. Most seek the voice of the living God but ignore the voice of his first-born son and hate the Children he sends to cross paths with them.

Hello, my friend!
I am God's only begotten son.
Will you believe in me?

THE PASSION OF THE NIGGA

HATE
PART II

Beatrice's father should've caught her before her mother did. Excuse the vulgarity but, I have freedom of speech so consider it spoken. Promises broken, her disregard for future consequences shitted up the polished veneer of our children's future. The fuck can you do with thirty pieces of silver in exchange for the soul of an American Aborigine mate and those of your own flesh and blood?

Stupid ass, Lying ass Bitch!

THE PASSION OF THE NIGGA

QUITTER

Among the Daughters of Zion are scores of quitters. Seeking life on their terms, they illuminate fashionable candles in the windowsills of Babylon and let them burn with hopes of finding a way out the shit they're in.

The Daughter of Doom once said, "A woman won't leave one nest until she's found another nest to go to." But, when the 33 and the 48 were leveraged against her, she let a Nigga know that another nest had been found and it was time to go. So, I opened the window to the wilderness and let her go. And, into the forewinds, she flew; failing to see that the 33 and the 48 showed her the way.

THE PASSION OF THE NIGGA

THE FIELD

Having seen the whole of Babylon in its magnificence, I am considered crazy by those who're either too blind or too deaf to perceive anything beyond the socio-economic and religious enclosure into which they have been set loose to sheepishly graze.

Behold, the Field.

THE PASSION OF THE NIGGA

GIVE IT BACK

Give it back. That first day I ever saw you. That first smile. That first "hello" and first "goodbye." That first "Happy Valentine's Day." That first kiss. That first argument. That first visit to Ward. That first introduction to my parents That first "I do." That first "I'll never leave your side"

Give me back that first "Where you go I go." That first day coming home to find you waiting. That first "I'm Pregnant." The Second and Third "I'm pregnant." Those emails begging you to be my wife. Those emails pleading with you to join me in the Delta.

Give me back the tears shed for you. Give me back all the kind words wasted on you. Give me back 16 years of my life.

Give it all back.

THE PASSION OF THE NIGGA

THE PASSION OF THE NIGGA

WHEN I WAS A SLAVE

I can remember when I was a slave; how the days brought frustration and anger my way and the nights were restless. I remember when my idea of freedom was remembering Martin Luther King Jr. or Malcolm X, or anyone in the pantheon of the so-called black heroes of America in the shortest month of the year - Freedom.

I can remember White women clutching their purses whenever I was near. I can remember the fear. I can remember feeling the stares and hearing the whispers as the only spec of color at Stingley Elementary School.

I can remember trying hard to fit in; fighting day and night to belong. I can remember how long I used to pretend to be

what I wasn't when I knew what I was. I remember caring what others thought. I can remember when Bob encouraged me to emancipate myself from mental slavery.

As I lifted my eyes to survey my surroundings, I saw Babylon and Babylon saw me. I can remember when I gained eyes to see. When Babylon found out that I had found myself, she hurriedly withdrew from me and took everything I loved. Then, she showed me the door.

INHALE EXHALE

The Delta flies by in hues of red and blue as Highway 61 winds North. Tightly rolled Ben Franklin between my lips, I dip deep into the Delta and inhale deeply of the Cush as Sunday's daughter sings songs of Zion.

No more fears of the darker days when Ms. Jane's daughter enacted the Meritorious Manumission Act of 2010 to erase me from my God-given rights as a father. No more thoughts of the days when Big Red used me to escape the Delta either.

The Southern landscape blurs into a tapestry of vibrant colors as I exhaled, and Highway 61 welcomes me to Tennessee. "Will I ever see the Brothers and Chocolate Chinese

again?" I ask myself as I draw long on that Franklin before exhaling the darkness that besets me.

Smoke spiraling into the Ether, I expel demons as Memphis rises on the distant horizon. As it rises, I declared to myself, "I made it." The game of life, I played it. The prayer of King David, I prayed it. The price for being a "Strong Black Man", most are too broke to remit it, but a Nigga was nailed to the walls of Madison County Family Court for the sins of Ms. Jane's daughter.

Destroying innocent children with hysteria, she gave a good life away for the approval of a Cracka whose Granddaddy lynched John Wilson on the steps on the same courthouse she used against me. As I rise above the Delta, I inhale and gaze back through the rearview to reconcile the future with my past.

THE PASSION OF THE NIGGA

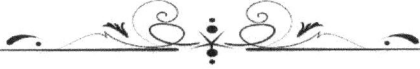

WHAT IF

What if you woke up and realized that everything you've been taught was a well-rehearsed lie? What if you learned that your very existence was a commodity that was traded on the auction blocks of the global directive? What if you realized that your birth, life, and eventual death bore a price tag?

How would you feel if you learned that you were raised, fed and bred according to an involuntary control matrix designed to engineer the perfect slave; a dumbly obedient slave that marched to the beat of a collective drum?

What if you lifted your eyes and witnessed the lie as its magnificent brilliance manifested itself in sights and sounds that were

easy for you to accept because you were conditioned to accept them?

What if you realized that the God you love is the Devil you hate? What if you were the only one that was awake? What if the identities of those who rule this world were revealed? What if "reality" one day became real?

THE PASSION OF THE NIGGA

TUESDAY

Tuesday changed my life. It greeted me with a letter from Ms. Jane's daughter delivered by a Washington County Deputy Sheriff. Ms. Jane's daughter lied on me! She lied on us!

For the world to see, she embodied the persona of a terrified and mentally abused Black woman oppressed by yet another tyrannical Black man.

In her wake, nothing remained but a shell of a man who was crucified for the sins of a Bitch who was too weak to "woman up" and be the wife she promised to be.

Nothing remained but a shell of a son silently crying out to his Momma because words escaped him. Nothing remained but a

shell of the chosen crying out to the God of Israel for salvation, but his prayers were unheard. I pined anxiously with thoughts of killing myself.

There I was, in the Delta, with nothing but thoughts of drinking until I couldn't feel the pain anymore. There I was, in the Delta, an outcast. There I was, in the Delta, a villain for being God's only begotten son. There I was, in the Delta, persecuted by Ms. Jane's daughter and the ones I needed the most. There I was, in Mississippi.

It all began on Tuesday.

THE PASSION OF THE NIGGA

AWAKENED LIABILITY

After being betrayed for thirty pieces of silver, my persona was brought before Pontius Pilate and tried for sins against the Matriarchy. Dragged from abstention, I was cast before the families who cried out in one voice, "Crucify him!" And, it was so.

Dead to the Law, my Corpus was laid before Ms. Jane's daughter who wept in the presence of my death before telling my sons I left them forever.

I was as far from my family, in life, as I would be in death. I was separated from my homeworld. I was distanced from my family. There, in the darkness, no one thought of me. No one pondered the condition of the slain. No

one looked upon my descent into the abyss with compassion.

I was lost to myself.

NATURAL SELECTION

Take away EBT, TANF, SECTION 8, CHILD SUPPORT and SSI, and natural selection will occur. The Modern Matriarch will crumble under the pressure.

For, she is a paper elephant infused with the blood of social Entitlement. Yea, even the advent of her Matriarchy is at the behest of enemies of whom she regards as allies.

A Daughter of Babylon is she and her days are few. When the weight of supporting entitlements becomes too much for her husband to bear, he will pass legislation to cut her off. He will divorce her, and she will walk in the power of her femininity uncovered and alone.

THE PASSION OF THE NIGGA

THE PASSION OF THE NIGGA

REGARDING JOBS

Sunday's daughter arrived in the Delta after the abomination that set up desolation. With Baby-Baby in tow, she brought rays of symptomatic sunshine and conditional rhetoric that sounded good to Niggas with a sweet tooth.

Sweeping redbones from underneath beds and closets, her resume was padded to portray the kind of shit that would win a woman a man in ninety days or less. She tried her best to balance reality from fiction but often got caught up in the friction between Ms. Jane's daughter, that Nigga who beat her ass and that thick ass redbone I used to fuck named Quin.

But, the wilderness was Sunday's daughter's chance to begin again minus the domestic violence. The wilderness was her chance to escape the cycle created by her mother. The wilderness was her chance to distance herself from the whoredom of her past. The wilderness was her chance to reinvent herself. Sunday's daughter arrived in the Delta desperate for a job.

THE PASSION OF THE NIGGA

SUNRISE IN THE KEY OF D FLAT

Through dusty blinds, I anxiously watched as the darkness yields to the light. From dull hues of blue and gray, the skies burst radiantly into shades of auburn, orange, and red.

Greeting a new day, I accepted the Delta, once more, as recompense for a poorly planned life. As the sun rose, the darkness was pushed back as the secret I concealed opened her hazel eyes and asked, "Can I shower before I go?"

Daddy used to say, "Every lie will come to light one day." As the sunlight pierced the darkness of my humble abode, the lie I'd been concealing rose from bed, wrapped a sheet around her naked body and headed to the

bathroom. I am a Human Being less than I was raised to be.

THE PASSION OF THE NIGGA

THE DELTA

I awaken to the reality of my present condition. Regretting the circumstances that delivered me hence, I hold fast to the dwindling hope that I will, perhaps someday, escape. I exist in a 360-degree panorama of systematic madness; surrounded by men, who are now, what I was when I was a child.

For, they're aloof. They're asleep to the world. They're deaf to the call of enterprise and dumb to their immediate need to flee the fields upon which their ancestors slaved.

I see the Delta. I see its people, but they look upon my personage as something strange and unintelligible. I am an ignorant outsider because my words and actions are of Sun Tzu and Hagakure. Despite it all, we occupy

different dimensions within the same living space.

An alien, I study them as they scurry about in what seems a pointless expression of life without living. They're the epitome of the stereotypic "Black Nigger," save a few who have, for some unforeseen reason, become more than the environment from which they emerged.

I hate this place with a kind of fanatical disdain that inspires desperate acts of self-liberation. The largest prison mine eyes have ever seen, it relegates all within to a kind of life that is best described as a waste of the Christian God's time. It is in this place that I've come to reside.

THE PASSION OF THE NIGGA

PAPER AIRPLANES

When I was young, I had a thing with paper airplanes. I used to beg my father to make one for me nearly every day. We used to fly them all over the house.

One day, my father flew one and almost hit me in the eye. After checking to see if I was alright, he told me, "Paper is dangerous young Judah. One sheet of paper can destroy an entire household. It can level nations. One sheet of paper can change the course of history."

I was confused. He saw I was confused and told me, "Getting hit in the eye with a paper airplane is nothing compared to what paper can do. Be careful my son and don't tell your Momma I almost blinded you".

THE PASSION OF THE NIGGA

I didn't understand my father's words until February 2nd, 2010 when a Sheriff's Deputy knocked on my office door and said, "Judah Weissman, you've been served." I opened the paper and discovered that my entire world was destroyed.

THE PASSION OF THE NIGGA

FRECKLED FACE

Quin's silhouette rhythmically rolled as moonlight shined through the shades, projecting her striped image against a distant bedroom wall. No barrier between she and I, she began to cry as her thrusts hastened and she began to quiver.

"I love you," she whispered as I thought back to that broken debit card reader and wondered if, by chance, its malfunction was cosmic destiny. Staring up into her freckled face, I watched as pleasure gripped her entire being distancing her from the harsh reality of the Delta.

As her love began to crescendo, her heaven beckoned to me and into her was poured my frustration, my need for love and

my   desire   for   my   wife.   Quin   was, simultaneously, a gift from the God of Israel or the Devil of Humanity. Only, I was too fucked up to rightly divide the two.

.

THE PASSION OF THE NIGGA

BEFORE

Lost in the starry magnificence of the Milky Way, I laid on my back, beneath the heavens, and wondered what my life would have been like before being sent to Alabama.

I wonder why my father never told me what he was planning for my future and why my mother never warned me. I wonder where my brother's "brotherly love" was when he signed my name instead of his.

I wonder how my life would've been different had I been allowed to choose my fate. I wonder how my life would've been had I been allowed to meet life, there, in Washington, D.C. instead of Tuscaloosa, Alabama.

How would my life had been different if I never met Ms. Jane's daughter? Would I have been blessed to watch my oldest son grow into a man? Would I have chosen Cassandra instead of someone else?

If I had been allowed to forge my path through the universe, would I have suffered so many failed relationships? Lost in the starry magnificence of the Milky Way, I laid on my back and wondered what my life would've been like before everything changed.

THE PASSION OF THE NIGGA

WE

We are Doctors and Attorneys, Artists and Pimps. We are Housewives, Husbands, Children and High School Dropouts. We are blind politicians and substitute Civil Rights activists.

We quarrel over a few insignificant things while our adversaries orchestrate our demise. We are unaware of the war being waged against us. We are quick to become victims and even quicker to shrink from responsibilities. We have destroyed part of our future for temporary gratification.

We take trophy White women as symbols of our success and leave our Queens alone to rear the children we left behind. We

have spent decades educating ourselves but are still ignorant. We are a team player.

We accept what we are told without question. We are afraid of rocking the boat. We seek the White man's approval but berate our Father's words of wisdom. We made it to where we are because we are better than our sister or mother or husband's ex-wife.

We are sick and do not know it. We are blind. We are still demanding the same rights Martin, Medgar, and Malcolm demanded. We are living the "American Dream."

KARMIC VALUE

What is the Karmic value of Trust? What is the Universe's answer to crimes against a Child of God? What is a mother's love that she can't see the Cancer that gnaws away at her child's bones?

What is the Karmic value of finding that one soul mate, losing that one soul mate, then piecing together the remnants of that soul mate with parts of broken women weathered by rape, molestation, child abduction, and post-traumatic stress disorder? What is the Karmic value of Polygyny in lieu of a deferred eternity?

As I contemplate my cosmic fate, I'd be remised if I failed to speak the obvious; if only the obvious could be expressed in words

that could be understood by the hearts and souls of others; the hearts and souls of mothers who've had sons who needed them but were too filled with pride to simply cry.

I take hold of my fate with a certain type of honest stoicism and gaze deeply into the murky expanses of an uncertain future as Karma haughtily gazes back.

THE PASSION OF THE NIGGA

BEATRICE

The worse person in the world is a woman named Beatrice. No names have been changed to conceal the identity of the innocent because the innocent are children. Defeated and destroyed, they long for their father's embrace but have been sentenced to a life that best suits that Bitch.

Momma used to always say, "That's the mother of your children." Daddy raised me to respect women, but they also raised me, to tell the truth, and shame the Devil. So, I am calling it like it is.

Beatrice is a Bitch! She's a crazy ass neurotic Bitch who would rather destroy entire futures than simply say, "I made a mistake" or

"It was me, not you." A female dog is what she is. It was how she was raised.

No child is born into this world with preconceived personality traits. Beatrice is a Bitch because her Momma raised her to be one and her Pappy let that shit happen.

My Nigga, Colossus, thinks she will sue me for putting her stupid ass out there like this, but I don't give a fuck. Let her try. She can't take anything from a Nigga who's lost it all. She can't kill a Nigga that's already dead. And, she can't hurt a Nigga that ain't got no mother fucking feelings.

THE PASSION OF THE NIGGA

VIRTUE AND THE WOMAN

A virtuous woman and a clamorous woman cannot inhabit the same body. It would cause a singularity in the space-time continuum. Bitch would straight shut down and start babbling incoherent bullshit like, "I'm the same as you," and "disagreeing with me is a form of verbal abuse."

When you begin to hear this shit, know that your Daughter of Zion is malfunctioning. Send her ass back to the manufacturer and get another one. Dealing with her after she has become a singularity is dangerous to you and a danger to the children she will take from you and use as weapons against you.

She may lie in Court. She may say you're a bigamist or that you're teaching your

children to shoot at police officers. She may send your children on a retreat for young men who don't have positive male role models in their lives.

She may lie to your parents, and they may believe her because they don't know you anymore. She may reduce you to abject poverty because she never wanted you but always needed your mother fucking money. "Virtue" spoils, in time, and falls to the ground. Its seeds are seeds of deceit and scorn.

THE PASSION OF THE NIGGA

A WORD ABOUT THE "N" WORD
HTTP://WWW.RACEANDHISTORY.COM

"Nigger" has been used to degrade Blacks in America, but its origin has been misinterpreted by the ignorant and used as tools of mis-education, racial and self-degradation. "Nigger" did not realize its roots in the Latin language. It found its origins long before Latin or the advent of the Dutchman in Africa.

It is said that "Nigger" is a derivative of Niger, which means "Black." This is what Eastern and Western European scholars offer as an explanation. It's true and actual meaning is god or teacher. Do the research and see. It is a sacred name that was taken and desecrated by our captors during legalized chattel slavery. The word's origin is Egyptian.

THE PASSION OF THE NIGGA

You will find that Nigger (N-G-R) is but one of an extensive list of sacred words that begin with "N" which originate from our motherland. When I was young, I cringed at the sound of that word. But, as I grew and learned, I realized that, just as everything originated in Africa, so did the term, Nigger.

Now, here is the question. Shall we move towards the vain pursuit of seeking the eradication of a word whose true meaning has been steeped in ignorant cultural acceptance, misunderstanding, and hatred? Or, shall we educate our youth of its true and everlasting meaning?

THE PASSION OF THE NIGGA

EPILOGUE
THE BIRTH OF THE PASSION

"The Passion of the Nigga" was born out of pain and suffering. I have, at times, attempted to determine exactly when my Passion began. At best, it began the moment I met Cassandra, on a bus in Washington DC, while attending Howard University.

She was the most beautiful woman I had ever seen in all of creation. Have you ever felt such an attraction to someone that just being close to them seemed electrical? This was the case with Cassandra.

Shortly after our initial meeting, we began dating. Before long, we were going steady. Shortly afterward, we began making plans for a life together. I knew, from the beginning, that she was my soulmate. I knew,

from the beginning, that I had spent several lifetimes looking for her. At that moment, on that bus, in Washington DC, I found her.

Life with her was beautiful. Looking back, I believe this was the happiest time of my life. I was everything to her. She was everything to me. We planned to get married after graduation.

On Thanksgiving, in 1993, my father announced that I would be attending the University of Alabama in Tuscaloosa. This was news to me because I was still an enrolled student at Howard University.

While there, my family and I crowded into the car and drove the half an hour trip from Birmingham to Tuscaloosa to view the campus. I was confused.

I had not applied at the University of Alabama to the best of my knowledge. That being the case, I did not want to embarrass my father or cause him any disappointment.

THE PASSION OF THE NIGGA

As it turned out, my mother, my father and my younger brother facilitated my withdrawal from Howard University without my knowledge. When I returned to DC to finish out the rest of that semester, I learned that Cassandra was four months pregnant. She told me after I told her I would be leaving Washington D.C. She was heartbroken.

I remember how she cried. Her tears were almost biblical; carrying a certain level of woe that I had never seen before. She eventually gathered her composure. I remember how she looked at me and told me that she was proud of me. I was shocked that she was pregnant.

I promised her that we would always be together. I begged her to come to Alabama with me. She promised she would try to transfer to Tuscaloosa. She asked how long it would be before I left. I told her January. I remember she told me there were things she

needed to take care of. More than a few days passed.

I didn't think anything of it because it was Finals and she took her academics seriously. Two days before the winter break, she called and asked about Finals. I told her I did okay in some and great in others. She told me she had to talk to me about something serious because she may not be able to say it to me after I leave.

Honestly, I thought she was going to break up with me. I guessed that was okay. It wasn't the first time a woman broke up with me. Based on what you've read, it for damn sure wouldn't be the last. I waited for her in my apartment until she arrived.

When she arrived, I took her coat, hugged her, and I rubbed her stomach. She withdrew as a tear ran down her face. She told me her mother was a single mother. She said she remembered her mother begging her father for money. She told me she swore she would never be that woman; a woman like her mother.

Then, she told me she got an abortion. You can imagine the rush of emotions that flowed through me. I was speechless. I thought we were going to spend our life together. I thought we were both going to transfer to Tuscaloosa. There I was standing there; speechless.

She kept asking me to say something.
She kept begging me to hit her.

All I could utter was, "murderer." I called her a murderer. I told her that she deserved to burn in Hell for what she did. She collapsed to the floor and wept as I stood over her with dismay.

I couldn't even cry. She cried out to God to forgive her. I remember she begged me to forgive her. She promised to give me another son if I would forgive her.

THE PASSION OF THE NIGGA

A son!

I backed away, and she grabbed my leg and wrapped around it, begging me not to hate her. She kept chattering about being alone with a child. She didn't want to be like her mother. I had a son, and she took him from me.

I was as embarrassed as she was. I never really had a chance to tell Daddy or my Momma. She never gave me that chance.

I left for Alabama never to return to Washington DC. Cassandra took me to the train station and promised she would transfer the next semester. That never happened.

Her grades were not good enough. I never told my parents that I got Cassandra pregnant. I tried my best to forget her, but she kept calling me.

In time, her beauty, her smile, and her radiant eyes dimmed in my mind, and her egregious act of murder was suppressed in my subconscious. There was no need to tell Daddy about my mistake. At least then, he would be proud of me. A son should bring pride and honor to his father.

One day, I entered Mary B's restaurant and saw an old friend sitting across the table from a beautiful young lady. I approached my friend and asked her to come and speak with me for a moment. When she came over, I asked about the young lady sitting with her. She told me the young lady's name was Beatrice. She was Ms. Jane's daughter from Ward, Alabama. I asked her to introduce us, and she did.

Over the next 36 months, me and Ms. Jane's daughter dated and agreed to get married. Looking back, it was all bad. There were warning signs every other day which gave reasons for she and I not to get married.

No less than six months after the marriage, I wanted a divorce. I can remember calling Momma and telling her I wanted to divorce Ms. Jane's daughter and come home to start over. Momma told me that God didn't believe in divorce. No one in our family believed in divorce. I heeded my mother's words and decided to persevere.

Ms. Jane's daughter and I left Tuscaloosa Alabama and relocated to St. Louis Missouri. I got a job with a chemical company, and she went back to school and eventually graduated with a degree in education. We also gave birth to our first child. Her name was Victoria, but I called her "Chocolate Chinese" because she had Asian eyes.

A year after my daughter was born, I accepted a job with another company, in another state, and we moved to Atlanta Georgia. In Atlanta, we gave birth to two boys. But, with all the blessings; a new house, and new cars, something still felt wrong inside. Deep down, I knew what it was.

THE PASSION OF THE NIGGA

I shamefully admit there was nothing Ms. Jane's daughter could have done to please me. I mean, she tried. She wasn't perfect, but she tried. In retrospect, I don't think she was happy with me either.

Over time, marital dissatisfaction gave rise to our first petition for divorce. I remember standing with her and her attorney inside DeKalb County Courthouse. Her attorney told us once we entered the courtroom doors, it was over. Ms. Jane's daughter looked at me as if to say, "what next"? I thought about the children. We decided to stay together for them. It was the honorable thing to do. It brought honor to her family, and it brought honor to my family.

Soon afterward, I began to evolve; spiritually. I stopped eating pork and started reading the Bible. I stopped going to church and found out that I was an Israelite. As an Israelite, there were certain things I could do

that regular Christian folk couldn't. I began to worship with other Israelites. Ms. Jane's daughter and my children joined me. It all seemed great! What more could one ask for?

Some years later, after developing many close relationships with some of the sisters in the tabernacle, Ms. Jane's daughter asked what I thought about polygyny. I didn't know much about polygyny.

I began to research it and realized it was biblically permissible and was also a plausible means for fixing what needed to be fixed without further breaking what had already been broken.

I did not love Ms. Jane's daughter the way a man should naturally love his wife. I always felt, from the very first day, that she was tolerating me. Polygyny became a way for me to make a complete set of China. Ms. Jane's daughter introduced me to a dark-

skinned woman who lived in La Vergne Tennessee. Her name was Cindi.

Cindi began to spend weekends with us. She began to intertwine herself with the family in such a manner to demonstrate her intent to be part of the family.

Cindi and Ms. Jane's daughter got along well. They shopped together. They watched television shows together. They played UNO together. They studied Scripture together. They cooked together. Everything seemed fine.

One day, after work, I came home to find Ms. Jane's daughter and Cindi seated on the couch. They asked me to sit and listen. Ms. Jane's daughter began to explain that it was okay for Cindi and me to know one another in the biblical sense.

Looking back at that moment, I should've cursed them both out, asked Cindi to leave and disallowed Ms. Jane's daughter from

ever speaking to me about polygyny again. I listened.

Although Ms. Jane's daughter was very beautiful, there was femininity within Cindi that eclipsed Ms. Jane's daughter. It still felt kind of strange though. Even though it was biblically permissible, I had reservations.

On a Tuesday evening, I returned from work and discovered that Ms. Jane's daughter and our children had gone up the hill to visit my parents. Cindi came to greet me in the living room. She took me by the hand, into her room. That was the first time.

Sex between Cindi and I happened frequently. I began to realize that Cindi desired sexual attention over that of righteous cultivation. She did not want to develop a marital relationship outside of the bedroom. Because of her appetite, I was diminishing from Ms. Jane's daughter, and she began to become jealous.

Ms. Jane's daughter wrote a letter explaining how she never intended for Cindi to join our family. It was a complete reversal. In the letter, she told me she preferred a younger wife who would be more easily moldable.

If Cindi did not get what she wanted when she wanted it, she would have an emotional breakdown making the house a living hell. The children were watching. Something had to be done.

I asked her to leave and wrote her a bill of divorce. The damage between myself and Ms. Jane's daughter had already been done. Ms. Jane's daughter was already scorned. It was only a matter of time before the inevitable occurred.

Much later, I was offered a better job in Mississippi. Because the job started during our children's school year, I decided to go ahead of my family and establish a place for them so that, when the school year was over, they would join me. Ms. Jane's daughter never joined me.

I used to call her on the phone, begging and pleading with her to join me in Mississippi. I needed my wife. I used to beg her to come to Mississippi to be the wife she promised to be.

Eventually, she stopped answering my phone calls. I would only talk to my children every second Thursday of the month. Every second Thursday was payday, and she would let me talk to my daughter and sons then ask for money. Once or twice she would visit, arriving early Saturday morning and leaving late Saturday night. She never stayed.

Her absence made me angry. I used to leave angry voicemail messages for her. I used to send angry emails too. I was alone. In Mississippi, nobody thought like me. I didn't know if there were any Israelites anywhere.

THE PASSION OF THE NIGGA

I considered committing suicide nearly every day. Each day I thought about it, the act of doing it became more appealing. I bought a pistol for the sole purpose of blowing my brains out.

The day I decided to actually do it was the day I met Quin. That evening, I called her. That night, she was on top of me doing her business. She was morphine sent by the God of Israel or the Devil of Humanity to lessen the misery.

She liked to do certain things to escape the harsh reality of the Delta. I had a void that needed to be filled. Every Thursday evening, she would drop her daughter off at her sister's house before visiting. Every Friday morning, she would get showered, dressed and leave for work. She was a Nurse.

Quin lessened the pain of loneliness but gave wide breadth for guilt; a pleasurable guilt comparable to the rush a drug addict experiences seconds after the drug enters the bloodstream. With each visit, she made more and more love to me until our "friend time" became more to her than it was to me. Quin was falling in love.

I began to feel like a monster. I was a liar. I became guilt-ridden when I was with her as she began to intermingle her hopes and dreams for the future with fucking. Listening to her, I knew she was worth more than what I was taking from her. So, I broke it off. No more Quin. Though we remained friends, she and I agreed never to do the things we were doing again.

I spent the next two years by myself; still anguishing over Ms. Jane's daughter and my children. The divorce was in full tilt. I found myself spending all my free time in the Law section of the Cleveland City Public Library.

THE PASSION OF THE NIGGA

The correspondence between myself and her attorney was like a volleyball match. He would send documents. I would rebut them and counter his motions. This went on and on for almost two years.

I used to talk to a friend named Khadiatu on the phone every night. She used to keep me company and far from the freckled temptation of Quin who used to drop by every now and then to see if she could 'do' anything for 'me'. Khadiatu always had encouraging words. She had her issues too, so we would help each other. She reminded me of Cassandra.

Two years and eight months later, Khadiatu showed up in Mississippi. I couldn't believe it when I saw her. She was there, at my front door, like an African ghost. I was glad to see her. She brought her daughter Nisoni with her.

THE PASSION OF THE NIGGA

We spent months talking about relationships; her past relationships and my past relationships. She told me all her secrets. I told her about Quin and Cindi. We debated whether we needed to love one another to be married to one another. Sometime along the way, my son was born. I named him Oz.

Eventually, Ms. Jane's daughter used those angry voice messages and emails I sent her against me. Her cause for divorce was because I had an adult female and her minor child living with me. She made no mention of Cindi living with us in Alabama.

When Khadiatu came to live in Mississippi, Ms. Jane's daughter made her move. This time, because she seemed emotionally distraught, downtrodden and victimized by an angry Black man, everyone believed her and cosigned her actions. My family believed her too. I was destroyed.

THE PASSION OF THE NIGGA

I felt betrayed even though I thought the divorce was justified. I felt abandoned because Ms. Jane's daughter outright lied about everything. Then, she took my children from me. The damage that occurred may never be undone. I was in a dark place.

There didn't seem to be a way out. I did not love Khadiatu the way I loved Cassandra. I didn't love her the way I loved Ms. Jane's daughter. But, I loved Oz more than death and I wasn't going to cheat him the way I cheated "Chocolate Chinese" and the Brothers.

I began capturing my feelings in free verse. It was a sort of spiritual bloodletting. This is how "The Passion of the Nigga" was born.

THE PASSION OF THE NIGGA

AFTERWORD

I come in peace. It was not my intent to offend or dismay, but rather purge my inner-self. To know me is to know the elements that made me. As I march towards my inevitable destiny, the fate to which I've subjected myself has given rise to this Passion. It is life; life as seen through the eyes of a Son, a Father and a Husband whom the world called "Nigga."

THANK YOU FOR HEARING ME

THE PASSION OF THE NIGGA

C'EST FINI

COMING SOON:

The Passion of the Nigga: Book II.

PROJETIC JUSTICE

www.ingramcontent.com/pod-product-compliance
Lightning Source LLC
Chambersburg PA
CBHW060537210326
41519CB00014B/3249